THE MOON

Rebecca Woodbury, Ph.D., M.Ed.

Gravitas Publications Inc.

THE MOON

Illustrations: Janet Moneymaker

Copyright © 2025 by Rebecca Woodbury, Ph.D., M.Ed.

The Moon
ISBN 978-1-950415-40-3

Published by Gravitas Publications Inc.
Imprint: Real Science-4-Kids
www.gravitaspublications.com
www.realscience4kids.com

RS4K

Photo credits: Cover, Title Pg, & Above: NASA, Public Domain; P.5, 9, 11, 13, 19. Earth & Moon, NASA, Public Domain; P.7. Top left, Courtesy of r the Illinois State Museum, CC BY SA 3.0; Middle & Bottom, Public Domain; P.15. By spiriterror from Pixabay; P.17. Marc Arias on Unsplash;

If you look up at the night sky,
you might see the **Moon**.

The Moon is
so pretty.

The Moon is **spherical**, or ball-shaped, like Earth. But it is much smaller than Earth.

The Moon is made of rocks and minerals like Earth. But the Moon has very little air and no liquid water. It does have some water in the form of ice.

I don't think I could live on the Moon.

Rocks from the Moon

We sometimes see the Moon as a bright light in the sky. The Moon does not make its own light. Instead, light from the Sun bounces off the Moon and comes to Earth.

I wondered where the Moon gets its light.

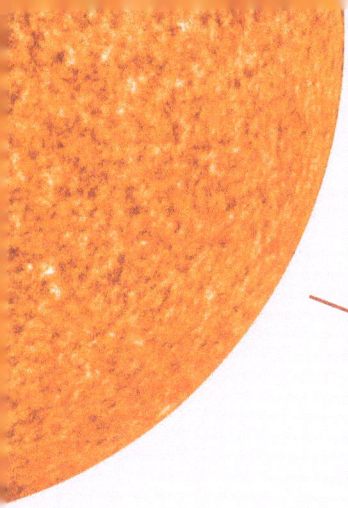

Sun

Light from the Sun

Moon

Earth

The Moon **orbits** Earth. The Moon completes one orbit every month. **Orbit** is the name for the curved path the Moon follows around Earth.

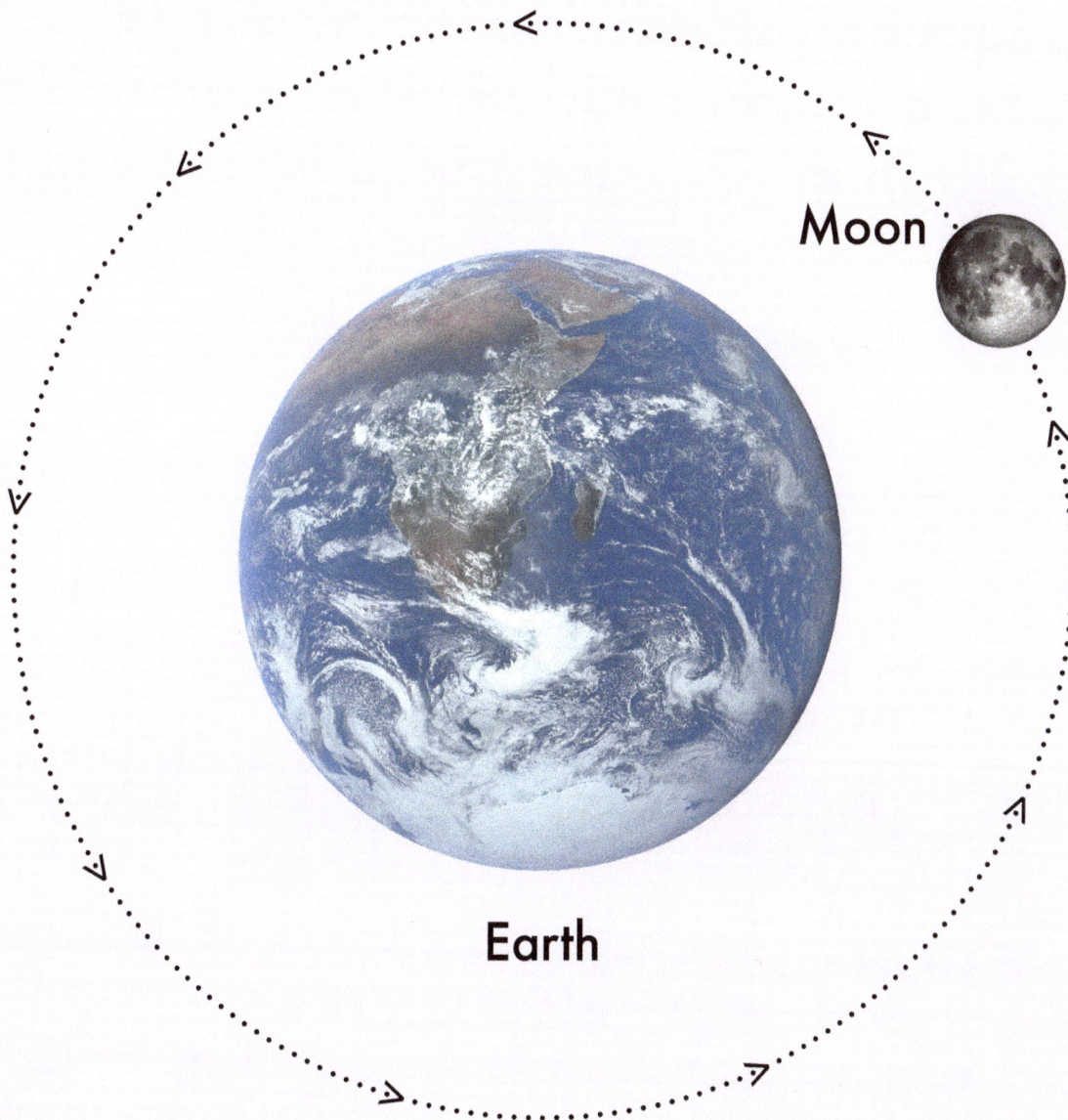

Moon

Earth

The Moon's orbit around Earth

During the month, the Moon looks different from day to day as it moves in its orbit around Earth.

The Moon can look completely **round** or **half-round** or **crescent-shaped**.

I have seen all of those.

Round   Half-round   Crescent

If you look closely at the Moon,
you can see light and dark patches
that may look like a face.

The light and dark patches are **craters** and **lava flows** from **volcanoes**.

A volcano and lava flow!

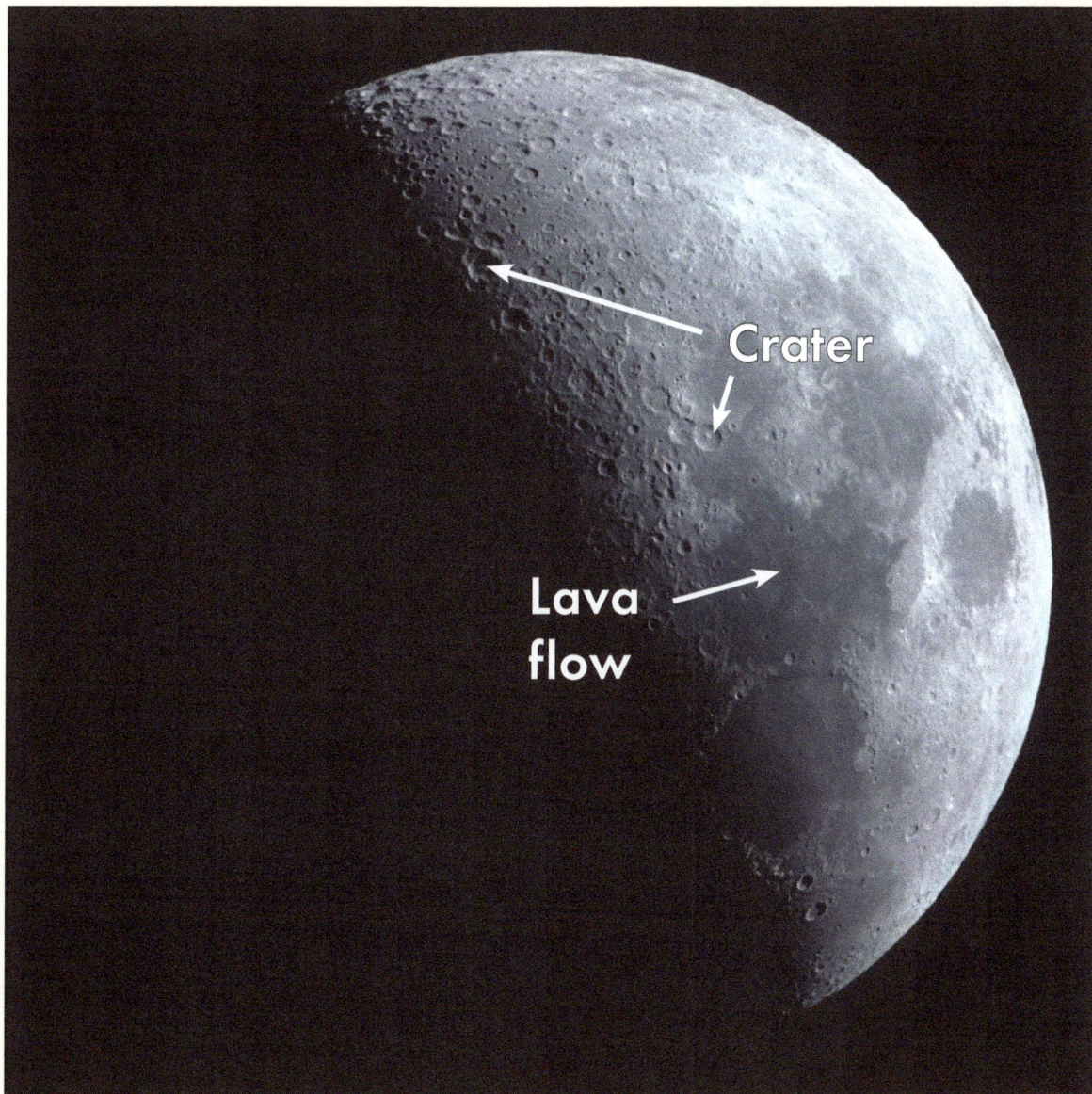

Crater

Lava flow

A **crater** is a bowl-shaped dent in the Moon's surface. A crater can form around the opening of a volcano. It can also form when another object in space hits the Moon.

That is a BIG dent.

The Moon and Earth each have **gravity.** Gravity is the force that keeps the Moon in its orbit around Earth. The Moon's gravity creates ocean **tides.**

Gravity is the force that holds everything to the surface of Earth.

High tide

Low tide

How to say science words

crater   (KRAY-tuhr)

crescent   (KREH-suhnt)

Earth   (ERTH)

gravity   (GRAA-vuh-tee)

lava flow   (LAH-vuh   FLOH)

liquid   (LIH-kwid)

mineral   (MIN-ruhl)

orbit   (AWR-buht)

science   (SIY-uhns)

space   (SPAYSS)

spherical   (SFIR-ih-kuhl)

tide   (TIYD)

volcano   (vahl-KAY-noh)

water   (WAW-tuhr)

www.ingramcontent.com/pod-product-compliance
Lightning Source LLC
Chambersburg PA
CBHW040153200326
41520CB00028B/7585